Embrace Your Existential Crisis

Breaking Down Barriers on Your Path to AI Confidence

Dina Alkhateeb

Raymond Vogel

ACKNOWLEDGMENTS

Thanks to all who patiently listen to us include "AI" in every sentence and allow us to take conversations into strange and sometimes outrageous new futurescapes.

Thanks to the extremely talented community of geeks, fanatics, and entrepreneurs keeping us all abreast of what's happening in the insanely fast-paced world of AI tools and technology development.

Last, thanks to those who have been so steadfast in encouraging and motivating us as we lean ever further into the frightful and thrilling changes of this fourth industrial revolution.

USE OF ARTIFICIAL INTELLIGENCE

Humans developed this product with the support of AI tools and technology. This included brainstorming topics and outlining, narrative inputs, graphic development, and iterative reviews to improve the final product. While the technology doesn't yet exist to "write me a book" (or not a good one, at least not at the time this book was written), we took every advantage we could to develop a high-quality product… as should you.

ABOUT FRONTIER ACADEMY

Frontier Academy was built by working professionals looking to harness AI technology to stay relevant in their industries. Our mission is to help individuals and teams stay competitive in rapidly evolving workplaces by leveraging the exponential productivity enhancements these tools offer.

Yes, we also offer training and consulting services.
If you're looking for more hands-on support, use the QR code below and give us a good way to contact you.

Learn more at FrontierAcademyLLC.com!

CONTENTS

THE DEATH OF IMPOSTER SYNDROME   1

INTRODUCTION   4

PART 1: WHAT EVEN IS AI?   7

Like a Really Smart Person   8

Where do They Keep the LLMs?   10

Are the Robots Forming an Army?   13

Why Does My Boss Care?   14

Okay, But is AI Coming for My Job?   15

What Can AI Do for Me?   16

PART 2: HOW DO I GET STARTED?   17

Dipping Your Toe In   18

Different Techniques for Different Days   19

Beyond the Frameworks   22

A Word of Caution   23

PART 3: WHAT CAN AI DO FOR ME AT WORK?   25

Step 1: Get Permission to Try   26

Step 2: Brainstorm AI-Worthy Challenges   28

Step 3: Make Your First Easy Button   30

Step 4: Communicate Your Value   33

PART 4: CAN AI ADD HOURS TO MY DAY?   37

Be an Easy Button Superstar   38

Unlock Easy Button Chains   40

Custom GPTs, Claude Projects, and Gemini Gems   42

Tailored Technologies   44

PART 5: HOW CAN I PROTECT MY CAREER?   **46**

You Might Be a Laggard If…   47

Get Out Over Your Skis   50

Close Your Eyes, Raise Your Hand   52

PART 6: WHERE IS ALL THIS HEADING?   **55**

Hard Right Turns   56

No Resume Needed   56

Interest-Based Culture   57

The Skills That'll Matter   58

Further Down the Rabbit Hole   59

PART 7: WHAT'S NEXT?   **61**

From Imposter Syndrome to AI Superstar   62

Beyond the Laggard   62

Continuing Your AI Journey   63

THANK YOU   **64**

ABOUT THE AUTHORS   **65**

THE DEATH OF IMPOSTER SYNDROME

Sarah, a newly hired campaign manager for a product release, begins her first day at the company with a mix of excitement and overwhelming apprehension. As she settles into her new office, her mind races with doubts and fears. "What if I'm not cut out for this job? What if I can't do this?" she thinks to herself. "I don't know the first thing about managing a product launch. What if I make a mistake and let everyone down?" Her hands tremble slightly as she logs onto her computer, feeling the weight of imposter syndrome bearing down on her.

She takes a deep breath and remembers the AI tools she had learned about during her onboarding. "I know AI can help me, but what if I don't know how to use it right? What if I rely on it too much, and it makes me look incompetent?" Despite her fears, Sarah knows embracing AI is crucial for success in her new role. She takes a leap of faith and activates her AI assistant.

The AI assistant greets Sarah and starts briefing her on the upcoming product launch. It presents her with a comprehensive, personalized plan, highlighting the similarities between this project and successful campaigns from the past. As the AI reviews each aspect of the plan with her, patiently answering her questions and addressing her concerns, Sarah begins to feel more at ease. She realizes that the AI is not here to replace her but rather to empower her and help her succeed.

As they collaborate, Sarah tentatively suggests a creative customer engagement concept. To her surprise, the AI assesses the idea and helps her refine it, incorporating the most promising elements into the plan. By midday, Sarah and her AI have generated a complete set of marketing materials, and Sarah's confidence grows as she sees the tangible results of their partnership.

Throughout the day, Sarah learns to trust the AI's guidance while still bringing a unique perspective to the table. She discovers that working with AI is not about relinquishing control. It's about blending its powerful capabilities to enhance her own skills and creativity.

As the day comes to a close, Sarah reflects on her journey from self-doubt to empowered confidence. She realizes that her initial fears and uncertainties were natural, and by embracing AI as a tool for growth and success, she has unlocked her true potential. Sarah now understands that the future of work lies in the collaboration between human ingenuity and artificial intelligence, and she is excited to be at the forefront of this transformative era.

~~~

In this near-future scenario, AI tools become the catalyst for eliminating imposter syndrome and letting us humans focus on those aspects of the job that got us excited about getting it in the first place. By empowering employees like Sarah to harness the power of AI, companies not only drive business success but also create a workplace where every individual can thrive, free from the limitations of self-doubt and equipped with the tools to achieve their full potential.

Sarah's story isn't some far-off fantasy. It's a glimpse of what's possible right now, thanks to the AI boom. No more second-guessing. No more "I hope I did this right!" moments. Just like Sarah, you can have a treasure trove of knowledge, automation, and expertise right at your fingertips.

Think about it: a world where imposter syndrome becomes a thing of the past, all because you've got your own AI sidekick to chase away those nagging doubts. Remember those times you felt thrown in the deep end, without a lifeline or know-how to tackle a new challenge? Those days are history.

Just imagine how different your first day on a new job could've been with your very own expert whispering in your ear 24/7. It's all about catching that AI wave that's crashing onto our shores. Don't be the one left standing on the beach, watching the tide rise. It's time to grab a board and ride!

INTRODUCTION

Ready to ride the AI wave? You've made the bold choice to dive in. You can mark that off your list. Check.

But now you might be thinking, "How do I actually start?" or even "I feel nervous about this."

We get it; you might be scared for a bunch of reasons. Don't worry, it's totally normal. Our brains are experts at building barriers, especially when we're facing something new. Any kind of change can flood us with dread, worries, and some pretty irrational fears. But hey, that's why this book exists - to be a personal guide through your very own existential crisis. We're going to tackle those irrational barriers head-on, the ones that might be holding you back from giving AI a real shot.

First off, it's important to understand that AI isn't going anywhere. This wave isn't about to peter out - we're only going to see AI become a bigger part of everyday life. Big investments are being made and better tools will continue coming out, and it's already a $20B+ industry.[1] It's a no-brainer, really. With AI tools, you don't have to learn everything from scratch anymore. You can get smarter, faster, and work more efficiently.[2] What manager

---

[1] https://www.hostinger.com/tutorials/ai-statistics

[2] https://mitsloan.mit.edu/ideas-made-to-matter/how-generative-ai-can-boost-highly-skilled-workers-productivity

wouldn't want employees like that on their team? We're saying all this not to scare you more but to motivate you.

Because here's the cold, hard truth: if you're not doing that already, you're falling behind. In fact, we'll go out on a limb here and say that if you're not embracing AI tools for your career growth and success, you're going to get left in the dust by those who are. Unfortunately, the kind of change you need doesn't happen overnight, so if we can nudge you just one step closer to being comfortable with this daunting new tool, we'll chalk that up as a win.

Also, remember Sarah from earlier? If you're itching to silence those nagging doubts and slay the imposter syndrome dragon once and for all, you'll need AI on your side. So, unless you're getting ready to retire, it's time to face those fears head-on and try to turn that anxiety into excitement. It's time to gear up and face those fears head-on with a digital ally that's got your back.

Let's talk about change for a minute. Dr. Everett Rogers, a statistics and sociology guru who wrote *Diffusion of Innovators*, was a big deal in the world of change management.[3] He dove deep into how different communities adopt innovations. Dr. Rogers sorted people into five categories when it comes to embracing change:

1. Innovators – These are the risk-takers, the adventurers willing to leap into new ideas and kickstart the whole adoption process.

2. Early Adopters – Think of them as the cool kids on the block. They've got the social clout to make an innovation look good and get others on board.

3. Early Majority – Cautious but curious, these folks are ready to try new things before the average Joe. They're the tipping point that turns an innovation into the next big thing.

4. Late Majority – The skeptics of the group. They need a lot of convincing, peer pressure, or maybe even economic necessity before they reluctantly join the party.

---

[3] Rogers, E. M. (2003). *Diffusion of Innovations* (5th ed.). Free Press.

5. Laggards – The "if it ain't broke, don't fix it" crowd. They'll cling to the old ways until there's literally no other option.

As you read through these, you might see a bit of yourself in one or more categories. If you're feeling like a Laggard right now, don't sweat it. Just by purchasing this book, you've got one foot outside the Laggard category. Our goal is that by the time you finish this book, you'll feel ready to join the ranks of the AI Early Adopters or even the Innovators. But more than that, we want you to see just how valuable AI can be in every nook and cranny of your life.

> **Pro Tip:** Use our free Laggard to Innovator Assessment GPT (Link under the "Thank you" section at the end of this book) to see where you fall and get ideas on how to improve.

Now, let's get one thing straight - this isn't your typical "How to Use AI" manual. We don't dive deep into prompt engineering or give you hundreds of pages of prompt ideas. Why? First off, we prefer to teach you to fish rather than give you what we caught, and we feel learning to communicate effectively with AI is much more important than having access to a library of pre-built prompts. And frankly, if you are actually hunting for a library of prompts for every situation, there are tons of books out there doing just that. Heck, there are even custom GPTs designed to help you write prompts!

Nope, we're taking a different approach here. Instead of force-feeding you prompts to copy-paste, we're going to help you conquer your fears and start to rewire your brain. We'll get you through those scary first keystrokes and teach you how to think about AI and even embrace it. We'll also give you some solid tips on ways to interact with AI in your job, your personal life, and everything in between.

How do we accomplish this? One step at a time. Each part of this book is designed to help you take a small step, face a new fear. In fact, at the start of each chapter are low-risk, high-reward activities that anyone can do, called "Quick Wins." Think of these as example training wheels – small steps that can lead to bigger journeys. We also recommend reading this book one chapter at a time. Take in a chapter, try out our suggestions, rinse, repeat. Your sense of vertigo will thank you.

PART 1: WHAT EVEN IS AI?

QUICK WIN — *Take 2 minutes right now to write down your biggest worry about using AI at work. Next, visit chatgpt.com (no account needed) and type: "What's a simple way to start addressing [insert concern]?"*

All right, before we lead you down the process of being an Early Adopter, we highly recommend you take a quick look at this part first. We're not promising to turn you into an AI expert overnight, but it's important that you have at least a solid grasp of the basics - what AI is and how it ticks. In particular, we're almost exclusively talking about Generative AI tools like Anthropic's Claude and OpenAI's ChatGPT. If you know what these are and roughly how they work, feel free to skip ahead to Part 2. No hard feelings!

Here's the deal: AI is evolving faster than a chameleon on a disco floor. This part of the book? It's written on the same shifting sands as the rest. The leaps in AI computing power are so mind-blowing, it's like trying to plan a picnic in a hurricane. So, heads up - you'll need to keep tweaking these strategies and tactics as often as you use them!

And this is just the tip of the iceberg. There's a whole ocean of books, podcasts, YouTube channels, and other resources out there if you want to dive deeper. We're giving you the 30,000-foot view here, just enough context to get you started. But don't worry, we'll toss in some links and recommendations in the sections below. So, if you get the itch to explore further, you'll know how to scratch it.

Like a Really Smart Person

What do we even mean by AI? Are we talking about robots or supercomputers or what?

First, it's important to note that we're not developers, we're AI enthusiasts. So we're zeroing in on the impacts and applications of the technology, not the nitty-gritty of how it works under the hood. While we recognize that many AI technologies exist beyond LLMs, including predictive analytics, recommendation systems, and adversarial networks, we won't get into those in this book.

Second, in this book, we are almost exclusively using AI to refer to multi-modal, Large Language Models (like OpenAI's ChatGPT or Anthropic's Claude). That's a lot to unpack, so bear with us. First, multi-modal simply means it can generate not only text but images, audio, video, or other types of content. The term "Large language model" (LLM) originated because of the size of the data used to train the system, but that's misleading at this point since these models extend so far beyond language. Natural language is exactly what it sounds like – you can talk to these LLMs as if they're another human. If you've heard of Natural Language Processing (NLP) or Machine Learning, these are types of AI that have been around for decades and the stepping stones to the LLMs we have today.

Next, let's talk about their size, because it's easy to imagine how LLMs are really big and really expensive. And it's true that creating an LLM takes millions of dollars, a lot of time, and powerful computing systems. But the version you're using online is surprisingly small. Small enough to run on a decent laptop without taking up too much memory. This is part of the magic that these AI companies have discovered. By smashing vast amounts of information into these probabilistic connections, AI models can quickly access and use large amounts of knowledge without requiring a ginormous amount of storage.

It's like they've found a way to distill the essence of human understanding into a concentrated form. It speaks to the ingenuity of AI researchers and developers, and it's one of the factors driving the current AI revolution. The ability to pack so much potential into a small space is what's enabling AI to be integrated into a wide range of applications and devices.

How it works is that these large AI models are fed, or "trained" by feeding it as much information as possible. The AI takes this information and identifies as many connections as possible. It only understands the information it was trained on – no more and no less. You know the predictive text on your smart phone? That's a good way to start thinking about it. That simple software essentially guesses (based on some probabilities and your phone use habits) what you might want your next word to be when typing a message. But here's another example to expand the idea.

Imagine you're at a coffee shop, and you overhear someone order a "venti, half-caf, soymilk latte with a pump of vanilla and extra foam." If you're someone who's spent a few hours of your life in a coffee shop (been "trained" on coffee making through extended exposure), you can probably guess what's going to happen next.

The barista will grab a big cup, pump some vanilla syrup in, add espresso (though maybe not as much as usual), pour in some steamed soy milk, and top it with extra foam. You can even guess that the customer might be particular about the temperature or ask for a sleeve for the hot cup.

Over time, you've built up this ability to predict the sequence of events from hundreds of observations of coffee being made, even if you've never made this exact drink. You've observed enough similar scenarios (been trained on data) to understand the typical flow of a coffee shop order (grasped the probabilistic connections).

Your brain has pieced together these experiences to form an intelligent guess of how coffee-making works in response to even a complex order. It can now make educated guesses about unfamiliar but related situations. This is similar to how generative AI models can make predictions based on patterns they've learned, even when faced with new, specific data they haven't seen before.

Interestingly enough, AI models have limits that are a lot like our own. For example, some LLMs are better than others at simple software tasks like information retrieval and telling the difference between fact and fiction. However, like humans, they're all great at telling stories and making creative leaps in understanding between disconnected ideas.

If you haven't heard of "hallucinating," it's an important term to understand, and it represents one of the more dangerous limitations of AI. Hallucinating is what happens when the model is more helpful than accurate, and what makes it dangerous is that the answer usually sounds so helpful. But it's not always based in reality, and it wants so badly to impress you that (unless you've given it permission to tell you "no") it will just tell you the story it thinks you most want to hear. Meaning, it can feed you the lies you want it to feed you, and you might like them.

It might be helpful to think of an LLM as someone who's great at guessing movie endings but terrible at remembering every line from a specific script. If you ask it to recite the script of a particular movie, it'll give you a script, all right. But it's very unlikely to match the original word-for-word. It's filling in gaps with what it thinks should be there, not what actually is.

This is why it's crucial to approach AI outputs with a bit of skepticism, especially when accuracy really matters. It's a powerful tool but not infallible. Always be ready to double-check its answers against reliable sources.

Where do They Keep the LLMs?

Taking you further down the rabbit hole here, there are a ton of different tools and technologies out there. According to the TAAFT (There's an AI For That) website, an average of more than 550 new AI tools are being released each month, touching roughly 4,800 different job functions, with many jobs having thousands of AIs tailored to related functions (Communications Manager, 5,900+

tools; Data Entry, 4,900+ tools; and the list goes on).[4]

To simplify what could easily be an entire book into itself, Frontier uses three rough categories of AI tools: AI Frontliners, Tailored Technologies, and Proprietary Power.

The AI Frontliners are the most popular and well-known tools on the market. They are available at a very low cost, have built-in data protection mechanisms and/or options, and are very likely to continue sticking around. Some big examples include OpenAI's ChatGPT and Anthropic's Claude. These tools are mostly used directly on their websites. They're also extremely low cost, if not free, to try out and learn how to use.

Tailored Technologies are custom-built AI tools designed for specific functions. Some of these tools use their own proprietary models, like Midjourney, which has its own LLM trained for image creation. Another big example is Meta AI, which answers questions for Facebook and Instagram users and has some similar multi-modal capabilities to ChatGPT. However, the vast majority of Tailored Technologies are products that run the models of others within their software.

In particular, many use models built by OpenAI and Anthropic to drive the results desired by their users. If you have Microsoft CoPilot on your MS Office subscription, this is a great example. Microsoft CoPilot is essentially ChatGPT tailored for use within MS Office products.[5] Users have an AI chatbot embedded in their MS Office products to streamline their work directly. Microsoft is also offering Copilot PCs that include ChatGPT built into their operating system for a faster and more seamless interface.[6]

Tailored Technologies have an extremely wide range of pricing, from free or very cheap (Midjourney starts at $10/month) to tens of thousands of dollars for more niche products. Because of the

---

[4] https://theresanaiforthat.com/

[5] https://www.microsoft.com/en-us/microsoft-365/business/copilot-for-microsoft-365

[6] https://blogs.microsoft.com/blog/2024/05/20/introducing-copilot-pcs/

continual release of these exciting products, we recommend treading carefully here in terms of how you spend your time and money. Smaller, tailored AI models may not be kept up to date, making the products less powerful over time compared to their competitors. Tools built on other models add value today, but future versions of AI Frontliners may easily perform the same functions.

Last, let's talk briefly about Proprietary Power. These are models trained and refined to do your (or your company's) bidding. For companies, there are services available to take a tool like GPT Enterprise and train it on your internal data to make it a super expert on your processes and needs.[7] This is a more costly endeavor (on the order of $250K and higher) but has the potential value to do much more tailored work. PwC famously spent upwards of $1B over three years on rolling out GPT Enterprise across their 75,000-person operation, including an "AI Factory" to provide training and generate enthusiasm.[8]

There are also individuals and small companies taking some of the extremely powerful, open-source LLMs and tailoring it for their needs. While this may be an increasingly accessible way to use AI in the future, right now it's mostly reserved for hard-core entrepreneurs and programmers looking to build AI into a core aspect of their products.

All that said, it's crucial to remember that this field is evolving at breakneck speed. These tools are powerful and relevant now, new tools are constantly emerging, existing ones are being updated, and the capabilities of AI are expanding almost daily.

This rapid pace of change means that staying up-to-date is not just beneficial. It's essential. To help you stay on top of these developments, we've created a dedicated page on our website where we regularly update our list of recommended AI tools (See our Thank You section at the end of the book).

---

[7] https://openai.com/index/introducing-chatgpt-enterprise/

[8] https://www.pwc.com/us/en/tech-effect/ai-analytics/generative-ai-impact-on-business.html

Are the Robots Forming an Army?

If your algorithms are giving you the scary side of AI, your news feeds might not be painting AI as humanity's new best friend. Nope, they're all about doom and gloom, serving up a buffet of AI hazards with a side of robo-pocalypse. So, we figure it's important to unpack at least a few of these fears.

First, let's address the big concern - the supposed existential threat of a robot apocalypse. With AI tools getting eerily close to mimicking humans, we're dipping into the "uncanny valley" - that uneasy feeling we get when something almost, but not quite, resembles a human.

You might hear terms like Artificial General Intelligence (AGI) and Artificial Super Intelligence (ASI) being tossed around. These refer to systems that could potentially surpass our understanding and pursue their own objectives. Some AI enthusiasts are even calculating the probability of doom, where AI decides humans are the problem and therefore expendable. Others hope for a utopia where AI solves all our problems.

While these discussions are intriguing, they're not particularly helpful to your daily life right now. Who cares about hypothetical situations that might happen? We have bills to pay. And here's the truth:

1. You can't do squat about it.

2. It's not gonna do you any favors right now - not at work, not at home, and definitely not in your wallet.

Even if AI truly poses an existential threat to humanity as we know it (and that's a big "if"), you've got two choices. You can spend your time worrying, or you can spend your time preparing. We firmly believe the latter is always more productive.

Moving on.

Why Does My Boss Care?

Your company's leadership team has likely been around the block a few times and seen software solutions blow up in their face enough times to be hesitant. And there are some legitimate worries with AI that are already in the news, including protecting hard-developed proprietary information and staying on the right side of the law. The risks and rules of this technology are very much still evolving, so they have good reason to be concerned.

That said, as they get smarter on how LLMs are used, in particular the big-name tools like ChatGPT, the risks go down pretty quickly. If you're about to have a discussion on this topic, or you're worried about these things yourself, here are some simple steps that can help keep you and your company protected:

1. Use a private or protected LLM subscription that includes terms that guarantee your data will not be used in training future models. It may cost a few extra dollars, but it offers better security.

2. Develop a habit of reviewing terms and conditions for any AI tools. Key questions to ask: *Will your data be used to train future models? Will you have a secure, proprietary workspace? Is there a way to opt out of data sharing?*

3. Be generally cautious with personal data. If you aren't sure your data will be protected, don't provide the tool any information you wouldn't be comfortable giving to a stranger.

Your boss may also be worried about AI-based automation getting outdated before the benefits make the effort worthwhile. They can immediately see the future: the team stops what they're doing and spends a ton of time and energy building some tool for a decent improvement, only to find out the technology has updated again. Nah, they'd rather just wait it out, let it mature. Unfortunately, there's some truth here. New AI tools are being released and updated daily. If you invest too much time re-engineering your work processes around today's technology, you might find yourself starting over before you even finish.

In this case, our recommendation is for everyone to stay flexible and adaptable. Instead of getting too attached to any single tool or process, focus on developing a mindset that can quickly incorporate new technologies as they emerge. The key is to stay curious rather than build up a bureaucracy that breaks when new changes come.

Also, keep an eye on emerging trends and be ready to adjust your strategies accordingly. This approach will help you stay ahead of the curve and make the most of AI's evolving capabilities.

Okay, But is AI Coming for My Job?

Let's talk about something that hits a lot closer to home: job loss. According to the SEO.ai report on 2024 employment job statistics, "30% of workers worldwide fear that AI might replace their job within the next three years," and that number is 74% in India.[9] And there's a reasonable basis for those fears. In particular, knowledge workers or "cognitive jobs" are being impacted the most. As an example, the Institute for Public Policy Research (IPPR)'s March 2024 study on the impacts of generative AI on work in the UK reported that the "here and now AI" is already capable of automating cognitive jobs with repeatable, consistent tasks (think jobs like administrative assistants, customer service, marketing, translators, etc.).[10]

We get it; this kind of data is scary and probably the reason you got this book. So, take a deep breath and listen. You can stay ahead of this curve; we know you can. Repeat that to yourself if you need to, as many times as it takes!

Let's break down this fear into something more manageable. The way we see it, your job security hinges on two important factors: (1) how quickly tools come to market that can do as good of a job as you and (2) how well you can use those tools to be more productive and efficient. If the point of this breakdown isn't already clear, we

---

[9] https://seo.ai/blog/ai-replacing-jobs-statistics

[10] https://ippr-org.files.svdcdn.com/production/Downloads/Transformed_by_AI_March24_2024-03-27-121003_kxis.pdf

think that those who resist using AI are the ones most at risk, and those who lean into it will have a significant and ongoing advantage.

So, don't panic. You're in the right place.

You also have some breathing room. This technology is increasingly expensive to improve, and there's still an adoption cycle that every industry will have to go through. It's definitely a shiny object that your bosses are looking at, but it can't do what they really need it to (yet), and it will be tougher to implement than they expect. In fact, if you're leading up a team or company yourself, we recommend you also pick up a copy of our other book, "Future-Proof Your Team," as it might help you start preparing your folks for the long-term culture change needed to stay ahead of your competition.

The smart move is to start thinking about how you can use AI to enhance your creative process, not replace it. The goal isn't to compete with AI, but to collaborate with it. Lean into it. And don't worry, we'll show you some ways how later in this book.

If it's not clear already, we think the biggest AI risk you face (that you can do something about) is falling behind.

What Can AI Do for Me?

This is really the best question to ask since it keeps you focused on the future and how to take advantage of what's happening.

To give you context, LLMs are extremely helpful in helping you tell the stories you need to tell. This might be the story of a vegetarian recipe to impress a date or the story to your boss of what that huge report really means. It might also be the visual, audio, or video story of what your imagination wants to show the world.

However, the truth is that this question is extremely personal. You'll have to answer it for yourself through a trial-and-error process that will likely go on indefinitely. Luckily, helping you start answering this question is woven into the fabric of this book. So add the phrase "What story do I need to tell?" to the back of your mind, and go bravely into the next chapter on getting started.

PART 2: HOW DO I GET STARTED?

That first part was a lot, we get it. And this next part, while fundamentally simple, will also be a lot because you'll get a glimpse of why we keep using adjectives like "powerful." Bear with us, though, because these first two chapters are when you get to experience your very own existential crisis.

Why a crisis? Isn't there an easier way? You could grow up in a world where AI is embedded in the fabric of everything you do, from your toaster knowing how you like your toast to your digital tutor helping you figure out a tough math problem. But if you're already out there in the workforce, trying to stay relevant and valuable, you need this crisis.

We recommend facing these changes head-on and turning those feelings of overwhelm and dread into a sense of urgency. This urgency will drive you to learn, to try, and to stay in control of your own destiny. Now, roll up your sleeves because it's time to get your hands dirty.

Dipping Your Toe In

If you don't already have a tool in mind, an easy place to start is where most people experienced generative AI for the first time: ChatGPT. Go visit chatgpt.com and either log in, sign up, or use it as a guest.

Next, type something. Anything. Ask it to write you a song about a Corgi with a bow tie or to tell you a story about a magical Corgi that saves a young child's life from a burning building. It doesn't even have to be Corgi-related; just try something.

Do it now. We can wait.

~~~

Fun, right!? And maybe a little intimidating?

As you keep asking the tool for more and more, you'll soon learn the two lessons that are at the core of nearly all debates about the practice uses of AI: (1) generative AI creates great content with very little effort, and (2) it's not as great as it seems, particularly if you're skilled in that area. As an example, if you write fiction for a living, you'll find that asking AI for a great story will leave you feeling very unsatisfied. If you're a graphic artist with a clear vision for an abstract image, you may find it difficult to create what you're looking for.

Try asking ChatGPT for something more challenging. Pick an area you know a lot about so you can try to find where the value of the tool starts to wane for you personally. It may not take you long.

Once you reach that threshold, that's where the fun stops and the work begins. Because you need to learn how to communicate with AI in the way(s) most useful to you. This is where the term "prompt engineering" comes from, which is basically the art of coaxing AI tools into giving you what you want. And it's daunting enough that jobs have been created, books written, consultants consulted, and tailored tools created—all to simplify this new work of learning how to communicate with these tools.

Our advice? Learn enough about this new art to extract value from the Frontliners. The return on investment for individuals is so outrageously high that there's really no downside to doing this. Plus, the rest of this chapter offers a lot of shortcuts to your learning curve, so you'll get there a lot faster than you think.

Different Techniques for Different Days

As your social media algorithms pick up on your interest in AI, you might start hearing about "prompt frameworks." Simply put, these are helpful ways to remember how to ask good questions from your LLM. Not surprisingly, they're at the core of prompt engineering because they help increase the consistency and quality of the response you get. From our perspective, we see them as a stepping stone that, over time, will help you refine your personal style of working with these amazing tools.

This section includes five Frontier Academy frameworks for you to try, with specific examples to test them out with. And yes, we recommend trying all of them, specifically for the learning reasons above. You can use our "Try This!" suggestions or make up your own. The more you try different methods, the better you'll get at extracting great results.

Rapid Fire

- **When To Use:** This technique is ideal for generating a large quantity of ideas quickly. It's particularly useful when you're stuck and need fresh perspectives or at the beginning of a creative process. The Rapid Fire Method helps overcome mental blocks and can lead to unexpected, innovative solutions.

- **Method:** (1) Clearly state the topic you need to brainstorm or the problem you want to solve. (2) Ask it to generate as many ideas as possible, encouraging wild and unconventional thoughts. (3) Ask it to build upon or combine some of the initial ideas. (4) Request that it identify the most promising or interesting ideas from the list.

- **Tips and Tricks:** Before starting, set a specific goal for your brainstorming session. After receiving a response, review the ideas critically and don't hesitate to ask for clarification or expansion on promising concepts. If the initial results aren't satisfactory, try rephrasing your request or specifying different criteria for the ideas.

Try This!

Use the Rapid Fire Method to brainstorm unique themes for a birthday party. Ask the AI to first generate ideas, then to combine some of them, and finally to pick the top 3 most exciting themes.

Precision Craft

- **When To Use:** Employ this method when you need a very specific, detailed output with particular requirements or constraints. It's great for creating structured content like plans or instructions. Use this when accuracy and adherence to specific guidelines are crucial.

- **Method:** (1) Define the exact output you need (e.g., a recipe, a workout plan), including an example or two if possible. (2) Specify all important details or requirements (e.g., ingredients to use, time constraints). (3) Describe the desired format for the information (e.g., step-by-step instructions, a table). (4) State any limitations or elements to avoid. (5) Request that it ask clarifying questions to ensure full understanding before proceeding.

- **Tips and Tricks:** Before using this method, create a detailed checklist of all required elements. After receiving the AI's response, carefully compare the output against your original specifications. If any aspects are missing or incorrect, don't hesitate to ask for revisions or clarifications.

Try This!

Use the Precision Craft Method to get a detailed, week-long meal plan. Specify any dietary requirements or preferred cuisines. Consider also asking it to include a shopping list.

Dialogue Dance

- **When To Use:** This technique is best for exploring a topic in-depth or getting personalized advice. It's particularly useful when you need to refine ideas or navigate complex subjects that require back-and-forth discussion. Use this method when you want to dig deeper into a topic and uncover nuanced insights.

- **Method:** (1) Pose a clear, open-ended question about your topic and offer to answer any clarification questions it has before responding. (2) Ask follow-up questions based on the response to dig deeper or clarify points. (3) Provide feedback on what's helpful and what's not in the responses. (4) Ask it to summarize key points or provide a conclusion after several exchanges.

- **Tips and Tricks:** Before starting, outline potential conversation paths or key questions you want to address. After each exchange, take a moment to reflect on the information received and formulate thoughtful follow-up questions. If the conversation veers off track, don't hesitate to redirect it to your main objectives.

Try This!

Use the Dialogue Dance Method to plan a garden. Start by asking about suitable plants for your climate, then dive into care instructions, layout ideas, and troubleshooting common issues through a series of follow-up questions.

Mind Meld

- **When To Use:** Use this approach when you need to analyze complex information, extract insights, or examine a problem from multiple angles. It's particularly effective for decision-making scenarios or when you need to synthesize diverse information into coherent conclusions. This method helps in gaining a comprehensive understanding of multifaceted issues.

- **Method:** (1) Provide comprehensive background information about your situation or problem. (2) Request a specific type of analysis or insight from the AI and offer to answer any questions it has about the background information or request. (3) Ask for alternative perspectives or interpretations of the information. (4) Request the AI to synthesize the information into key takeaways or recommendations.

- **Tips and Tricks:** Before using this method, gather all relevant data and information about your topic. After receiving the AI's analysis, critically evaluate the insights and cross-reference important points with reliable sources. If the initial analysis isn't comprehensive enough, consider breaking down your request into more specific questions or providing additional context.

Try This!

Use the Mind Meld Method to analyze and improve your sleep habits. Describe your current sleep routine, any issues you're experiencing (e.g., difficulty falling asleep, waking up tired), and your daily schedule. Ask the AI to analyze potential factors affecting your sleep quality, suggest improvements to your sleep hygiene, and provide alternative perspectives on addressing your sleep issues.

Beyond the Frameworks

Prompting frameworks are helpful tools in learning to communicate with LLMs, but you'll likely only use them for specific types of repeat tasks. For the rest of the time, when you need a quick explanation or need it to explain a document, you'll find yourself chatting more and more with the tools like they're helpful friends. Which is exactly what should happen.

To make all your conversations useful, below are a few additional tips and tricks that you don't have to try now but are definitely good to know.

- **Help it help you.** Remember, these tools can't read your mind (yet), and we humans rarely communicate with perfect clarity. When in doubt, tell it to ask you questions until you're both on the same page.

- **Give lots of context.** Whenever you can, be sure to give your tool example outputs, information you consider relevant, etc. Studies show that LLMs are exceedingly good at finding patterns in the content they're given but not always that great at following new rules.[11]

- **Use it as a prompt expert.** LLMs love to help you create prompts to use, so just explain what you're trying to do and ask it to give you some prompt options. It might save you some time and give you ideas on future prompts, too.

- **Think like a programmer.** Prompts are essentially tiny pieces of software written in your voice. Good software requires testing, updating, testing, and more updating. Effective prompts require the same thing, so feel free to play around with them until they give you the consistent results you're looking for.

A Word of Caution

One mistake almost everyone makes when starting to work with any Generative AI tool is to ask for the world. We want AI to do all 50 steps at once and just... do the whole project for us. In one prompt. When we see the power of a small prompt, we tend to project forward into what bigger prompts can do and start working on it. But, as you'll probably experience, that's when today's tools start to break down. Unfortunately, understanding lengthy, complex instructions is not one of their strong points.

---

[11] https://venturebeat.com/ai/llms-excel-at-inductive-reasoning-but-struggle-with-deductive-tasks-new-research-shows/

To keep your results consistent and accurate, start small. If you find the tool is still inconsistent, break your request into several small requests. Think of it like you're teaching an assistant, and you need them to crawl before they run. Each request gives them part of the context and experience they need for the real request coming later.

That said, the field of AI is rapidly evolving. The smarter models are much more likely to understand your complex prompts already. So stay curious, keep trying, and keep learning. The "language" of AI interaction is still being written, and you have the opportunity to help shape it!

PART 3: WHAT CAN AI DO FOR ME AT WORK?

All right, now that you've tested out a few simple prompting frameworks, it's time to apply that knowledge to your career. Remember Sarah from the start of this book? She integrated AI into her new job, making her daily duties easier and more efficient and essentially gaining a personal assistant.

As you find ways to embrace AI in your workday, you'll likely see some of those lingering work anxieties and uncertainties start to fade. First, because you can start to streamline or automate your more trivial tasks, giving you time to focus on the important stuff. Second, because you will get better and better at using this "expert in everything" by your side to help you get good at that important stuff. And last, because you'll look impressive to your boss and colleagues, who may not have caught onto the AI wave just yet.

This chapter walks you through a simple four-step process to start dabbling with generative AI in your work life.

Step 1: Get Permission to Try

If you work for a company that embraces change and innovation (or if your boss gave you this book), getting permission to use generative AI for work tasks might be as simple as asking. But let's face it, not all workplaces are created equal when it comes to adopting new tech. If your company or boss seems hesitant, try going through the steps below to see if you can help them see the light.

1. *Try to understand their concerns.* Put on your empathy hat and gently probe for insight. Are they worried about data security? Job displacement? The return on investment? Take a few notes and offer to do some research and get back to them.

2. *Do your homework.* Use some of your new AI prowess to input your notes into your personal account and request topics of research and other ideas on the best way to respond. Remember, some tools may give you what you want to hear, so be careful not to skip the "research" part of the process. In other words, make sure the results would pass a lie detector test before sharing with your leadership. Below is an example prompt on data security that includes research on Claude's terms and conditions.

PROMPT EXAMPLE (RESEARCH EMBEDDED):

The following comes from the Claude consumer terms and conditions: "We will not train our models on any Materials that are not publicly available, except in two circumstances: (1) If you provide Feedback to us (through the Services or otherwise) regarding any Materials, we may use that Feedback in accordance with Section 5 (Feedback). (2) If your Materials are flagged for trust and safety review, we may use or analyze those Materials to improve our ability to detect and enforce Acceptable Use Policy violations, including training models for use by our trust and safety team, consistent with Anthropic's safety mission."

Given this, can you briefly explain how Claude will help keep my work information secure if I use it in chats?

3. *Consider their communication preference.* Does your boss like to talk things through aloud? Get an email they can read at their leisure? Are they impressed by fancy reports and analysis? You know them best; the key is to meet them on their turf.

4. *Build your argument.* Pull together all your research and AI response notes into a single document. Then, ask your personal AI tool to help you present this information using your boss's preferred communication style. If they like to talk, you might need bullet-point reminders. If they prefer to read reports, you might need clear headers, an executive summary, and information to support each conclusion.

Pro Tip: If your boss hasn't read our book "Future-Proof Your Team," consider getting them a copy. It's easy to read, addresses most leadership concerns with using generative AI tools at work, and might open the door wider than you expect. Plus, there's a discount code at the end of this book under the "Thank You" section.

5. *Include the positive outcomes.* However you plan to communicate, and particularly if you think they think in terms of Return on Investment, you might want to include some information not just addressing their concerns but on the benefits of the tool. Below are a few studies you might want to look over (and even reference):

- https://mitsloan.mit.edu/ideas-made-to-matter/how-generative-ai-can-boost-highly-skilled-workers-productivity
- https://assets-c4akfrf5b4d3f4b7.z01.azurefd.net/assets/2024/05/2024_Work_Trend_Index_Annual_Report_663d4 5200a4ad.pdf
- https://www.hbs.edu/ris/Publication%20Files/24-013_d9b45b68-9e74-42d6-a1c6-c72fb70c7282.pdf

6. *Suggest a trial.* Suggesting a small trial period might be a good way to allow them to see some results without opening the company up to the bigger risks they're worried about. This also helps you present yourself as a forward-thinking problem solver, and that's valuable in any workplace.

As you work through this process, below are a few useful tidbits that might save you additional time in your research.

♦ The biggest data security concern is that you will give data to an LLM, and that data will be used to train future models. Look for terms and conditions (or Frequently Asked Questions) that discuss this particular issue and how to prevent it.

♦ Cost concerns usually relate to the big dollar figures used in the news and the magnitude of major AI implementation projects. The simple LLMs you want to try have an extremely low price tag, and they may even be free for limited usage.

♦ Generative AI tools aren't single-handedly replacing any jobs just yet. They're mostly being used to increase the speed and quality of results in jobs where menial tasks get in the way of adding real value.

Step 2: Brainstorm AI-Worthy Challenges

Alright, assuming you've got your permission to at least try some things, let's talk about identifying the ways AI might be most helpful in streamlining or even automating. If you've already pinpointed some ways to use AI, feel free to skip ahead.

To tackle this task, our recommendation is to look at the problem from multiple angles, not just within your role but for your team and your company as a whole. This will help you see the breadth of potential applications and may even help you understand better how your role can impact your company's competitiveness in general. Below are some actions you can take to accomplish this. Don't forget those best practices from Chapter 2, like having it ask you follow-up questions to get better responses!

1. *Start building a list.* Start building a list that includes challenges

and ways an LLM could help you accomplish or streamline them. As you go through each of the other actions below, keep adding to this list. Don't worry too much about duplicates and overlaps, your AI assistant can sort through that for you later.

2. *Analyze your resume and current position description.* Take a copy of your resume and job description and upload them to your generative AI of choice. Ask it to identify areas within your role that are (a) either challenging or trivial and (b) could be easier if you used an LLM to help you.

3. *Analyze other roles and responsibilities.* Following the same steps as action 1, analyze the job descriptions of your team members and even your manager/supervisor. In this case, be sure to ask how your role might make their roles easier. Since there is likely to be a heavy overlap between everyone's responsibilities, this may uncover additional challenges that an LLM could help you with that make a big difference to your boss and team.

4. *Analyze your guidelines or procedures.* If you follow any company-specific or even industry-specific processes, upload these documents (one at a time if they're large) in a new chat. Ask it to summarize the process and point out key areas that an LLM might be used to reduce the effort required to accomplish individual tasks. If you work with it for a while, see if you can get it to identify process efficiencies or gaps, areas where your position has the largest workload, and other potential process pain points or bottlenecks. As you iterate, remember to keep building your list of challenges that an LLM might be able to help streamline.

5. *Compare your position and resume with your company's vision, mission, or strategic plans.* Upload company-specific strategic statements or documents alongside your position description, then talk to your LLM about (a) challenges your company might have in achieving its goals, (b) the ways your role impacts those big-picture items, and (c) how an LLM

might be able to help you have an even bigger impact. If your company values proactive, "above and beyond" thinking, this is a great way to do exactly that.

6. *Play a "Guess my challenges" game.* Normally, giving an LLM plenty of context can be extremely helpful, but sometimes, you just want to see what it thinks of using a clean sheet of paper. Consider giving it your company's name, industry, and job title only, and have it guess what challenges your company and you might be having. Then, tell it which challenges were closest to right and have it guess again. Once you get bored of this, pick the challenges that are most relevant to you and ask how an LLM might be able to help you tackle them. You might be surprised by what it can come up with.

7. *Interview yourself for more ideas.* Ask your LLM to play the role of an investigator looking to uncover hidden challenges in your job and team that an LLM might be able to help streamline. Have it ask you clever, open-ended questions one at a time until it comes up with content that would add nicely to your list.

Remember, as you work through this process, keep updating your list of all the challenges and potential ways generative AI could help you tackle them. This list will be valuable in helping you work through Chapter 4, plus you'll need at least one good idea for Step 3 below.

Step 3: Make Your First Easy Button

In Chapter 4, we're going to recommend you start building a whole pile of Easy Buttons to make you immediately more productive. First, though, you have to go through the process of making your first one. So, let's get into it.

To make an Easy Button, it's helpful to understand how you prefer to think and work. In 2023, Harvard Business School identified two different ways of working with the technology and named them Centaurs and Cyborgs. Here's what they said: "One set of consultants acted as 'Centaurs,' like the mythical half-horse/half-human creature, dividing and delegating their solution-creation activities to the AI or to themselves. Another set acted more like 'Cyborgs,' completely integrating their task flow with the AI and continually interacting with the technology."[12]

In other words, centaurs tend to take more time to set up the tool and enable it to accomplish bigger tasks by itself. They spend more time iterating their prompts to make them create a consistent outcome that they can use over and over. Cyborgs, however, tend to use AI like a co-worker, mixing and matching small portions of tasks based on what feels comfortable or who can do it better.

Easy Buttons can absolutely work for both types of workers, but a Centaur's Easy Button may be significantly more complex and take more time to create. For this step, we recommend starting with a Cyborg-style Easy Button, because both types will help you, and you'll have plenty of time to build more complex solutions later.

Follow the simple 4-step approach below to accomplish this.

1. *Pick a simple challenge/solution combination.* Using your list from the previous section, find (or ask your LLM to help you find) a challenge that (1) happens pretty often in your job and (2) could be relatively simple for an LLM to support. We're guessing a few things already jumped out in your mind, but just pick a simple one for this exercise.

2. *Get more details on how an LLM can help.* Talk to your LLM about the challenge and solution combination. Brainstorm a list of ideas on how the LLM could specifically help. Expand on the ideas iteratively until you know what you want the LLM to do. Then, and this is the important part, ask it to act like a prompt engineer and create the prompt that will accomplish this for you consistently and accurately.

---

[12] Dell'Acqua, F., McFowland III, E., Mollick, E., Lifshitz-Assaf, H., Kellogg, K. C., Rajendran, S., Krayer, L., Candelon, F., & Lakhani, K. R. (2023). Navigating the jagged technological frontier: Field experimental evidence of the effects of AI on knowledge worker productivity and quality. Harvard Business School. https://www.hbs.edu/faculty/Pages/item.aspx?num=64700

3. *Test it, update it, and test it again.* In a new chat, test out your prompt using real text, data, or documents that might be a good reflection on what you normally would need it for. Even if it seems like it would be helpful enough for your purposes, test it again in yet a new chat—maybe do this a few times to be sure. And if something isn't quite right, ask AI what wasn't clear about the instruction and how to improve it. Be careful not to turn into a Centaur here (and lose yourself in the process of creating the perfect prompt), but also be sure to create something that will add value to you when you go to use it again.

Pro Tip: If your prompt seems complex or your LLM has a hard time following your instructions consistently, you might need more than one Easy Button for that task. Not only is there nothing wrong with this, but it's usually easier and faster to create two Easy Buttons than to make a single complex one work!

---

4. *Make the Easy Button... well, easy.* Copy and paste your final prompt(s) somewhere it will be handy for you. Consider a document that includes a link to your LLM, the prompt you use, and an example of what type of content to include. You know how you work best.

If you find yourself stuck in the quagmire of prompt creation and inconsistent or inappropriate results, below are some quick tips to help you get unstuck:

♦ Break down the challenge into smaller challenges, or break down the solution into smaller steps. Try asking your LLM for things like "Give me step 1 to start solving this problem."

♦ Have a good back-and-forth dialogue in a single chat. Each interaction builds on the last, so this can help your LLM better understand your objectives and gradually provide better answers to your questions.

♦ Ask for alternatives. For example, if you're working on an email

Easy Button and the tone is off, ask for example tones that might be applicable, then pick one that fits your goal best and try that. The words you use make a difference, so having choices will help you pick the right words.

♦ Give blunt feedback. If you don't like something about the response you're getting, say something. "I like the structure of the email, but you didn't include any good examples."

♦ When in doubt, give it more context. In the email example, you might need to give it a few example final emails that you liked and ask it to identify patterns that should be reflected in your prompts to create them.

Pro Tip: Use our free Easy Button Assistant GPT (Link under "Thank you" section at the end of this book) to help you build or troubleshoot the Easy Buttons you're trying to create.

Remember, you're still the expert in what you need. Generative AI tools can help you get there faster, but don't be shy about making the tool work for you. It's the combination of your unique insights and the tool's training, creativity, and speed that will generate the exceptional results you're trying for.

Step 4: Communicate Your Value

Now that you've danced with your AI partner and generated your first Easy Button, it's time to showcase the value it's bringing to your role and your organization. But here's the thing - projecting this value isn't always as straightforward as presenting a neat set of numbers. Let's explore how to effectively communicate the impact of your AI initiatives.

First, let's talk about how your organization "hears" value. Is more value placed on progress and process, or is it all about the numbers? Do they need to see immediate results, or are they comfortable with long-term shifts? Think about what your manager and their manager ask about first in a results-type meeting.

Also, consider that positive results look different for different professions. For example, an HR professional might create an Easy Button that simplifies their candidate reviews, while a marketing specialist might generate content more tailored to their target client. Both of these examples are difficult to quantify but have measurable impacts on their productivity and impact on the company. Here are a handful of example ways an Easy Button might add value:

- Faster review and synopsis of complex reports or data sets
- Higher quality writing clarity, coherence, and tone
- Better diversity of ideas for problem-solving or idea creation
- Faster topical research for unique needs
- More personalized report creation
- More personalized skill development and learning
- Faster integration of ideas into a single result

These qualitative improvements, while sometimes harder to measure, can be just as impactful as hard numbers. They often translate to long-term benefits that extend beyond immediate productivity gains. However, it's important to at least attempt to quantify value, even if you can only take it so far. Below is a simple approach for calculating and communicating the impact of your efforts to your leadership.

1. *Establish a "baseline timeline."* Estimate the time it would have taken to get to the same result. Include the additional research, time needed for a separate expert reviewer, time spent iterating the result, and whatever else it would have taken to achieve the quality and quantity of content your Easy Button helps you create. Use minutes or hours or days, whatever best fits your situation, but make sure they're countable.

2. *Document a "final result timeline."* Using the same format as your baseline timeline, identify the new times required for each aspect. For example, if an "expert review" portion is

accomplished by your LLM instead of another person, include the 30 seconds or so required there. If there are any new steps needed to use the Easy Button (like fact-checking), be sure to include these here.

3. *Do a side-by-side comparison.* Put the two sets of data next to each other in a table or narrative, whichever your supervisor will understand better. Be sure there is a clear alignment of tasks, even though the two processes may be notably different. This will have the added benefit of helping you think about ways to improve future Easy Buttons.

4. *Summarize the results at the top.* Before the detailed comparison, we always recommend including a Bottom-Line statement at the top of your communication or report. This should include both the qualitative value and the total quantitative improvement (in numbers) justified by your side-by-side comparison.

5. *Include a recommendation at the bottom.* Since you've been thinking about this Easy Button a lot, you probably have more you'll want to say. Consider including a simple "going forward" type recommendation that will show you're thinking strategically. It might be as simple as your plan to create more Easy Buttons, an offer to share your Easy Button with other teams, or even an offer for you to train others to build their own!

The other edge of this sword might (fortunately or unfortunately) be that your leadership is so excited that new expectations are set, both for productivity and communication of value. This is a potentially dangerous slope to be on, because the effort required to communicate value might start to undermine the actual value itself. We recommend talking to them directly about the time it took to document and calculate the value and working with them on ways to reasonably track value delivery over time.

Here are a few examples you might discuss:

♦   Periodic progress reports to highlight major AI-related achievements

- "Round-the-room" team presentations on Easy Buttons created and the benefits of them
- Sharing positive feedback from clients or colleagues when it relates to AI-assisted content
- Simplified tracking of baseline and Easy Button timelines for small improvements based only on gut feel (e.g., "took me about 4 hours before, now usually about 20 minutes")
- Sharing good "before and after" example products when there is a clear comparison
- Schedule feedback sessions with your manager to talk aloud through your latest creations and their impacts
- Planned brainstorming sessions with your manager or team to identify additional ways AI tools could help your team

Your goal is to demonstrate that AI isn't just a shiny new toy, but a powerful tool that's delivering tangible benefits to your work and the company as a whole. By presenting clear, measurable results and a vision for future improvements, you're positioning yourself as an innovative problem-solver who's driving the company forward.

If this feels daunting, don't worry, our next chapter has a ton of advice on ways to ramp up your value over time. You're doing great, just keep reading!

PART 4: CAN AI ADD HOURS TO MY DAY?

In today's fast-paced work environment, time is our most precious resource. What if you could magically add hours to your day?

In the last two chapters, we gave you the building blocks for working with generative AI tools and starting to add value at work. In this chapter, we continue our focus on "at work" since that's where most of our clients are concerned, but these best practices are applicable to making your time at home more productive as well. In particular, we dive deeper into ways AI can become your personal time multiplier to streamline your tasks and boost your productivity. This includes starting to create your own library of Easy Buttons, stringing them together to create Easy Button Chains, and even creating your own simple automation tools.

This is where the magic will really start to happen, creating real-time savings so you'll feel more and more like hours are being added to each day. We fully expect you to refer back to this chapter often on your journey.

Be an Easy Button Superstar

Becoming an "Easy Button Superstar" means systematically starting to build a portfolio of Easy Buttons that are at your disposal to streamline all those mundane, repetitive tasks that eat up your time and distract you from your most fun and interesting work. By integrating these tools into your workflow, you're not just saving time, you're making your work more enjoyable.

How do you build your portfolio?

There might be an Easy Button to create Easy Buttons one day (or you can create one yourself later in this chapter!), but for now, your best bet is to build them one at a time. To help you start with the goal in mind, below are some simple examples of what Easy Buttons might look like (minus the applicable attachments and company-specific context).

♦ *Create a Compelling Product Launch Email.* Write an engaging email to announce our new software product. Include the key features, benefits for users, and a clear call to action. The tone should be professional yet exciting, and the email should be no longer than 300 words.

♦ *Develop a 30-day Content Plan for Social Media.* Generate a month-long content plan for our company's social media accounts. Include post ideas for LinkedIn, Twitter, and Instagram, focusing on our sustainable manufacturing practices. Provide a mix of educational content, behind-the-scenes glimpses, and customer testimonials. Suggest optimal posting times and hashtags.

♦ *Summarize a Technical Research Paper.* Condense a 20-page academic paper on quantum computing advancements into a 1-page summary. Highlight the key findings, methodologies used, and potential real-world applications. The summary should be understandable to a general audience with basic scientific knowledge.

♦ *Draft a Customer Apology Letter.* Compose a sincere apology letter to a valued customer who experienced a significant delay in their order delivery. Acknowledge the inconvenience caused, explain the steps being taken to prevent future issues and offer appropriate compensation. The tone should be empathetic and solution-oriented.

♦ *Design a Problem-Solving Workshop Outline.* Create an outline for a two-hour workshop on creative problem-solving techniques for a team of software developers. Include icebreaker activities, an overview of key problem-solving methods, hands-on exercises, and a debrief session. Provide time allocations for each section and list any needed materials.

That said, these are just example ideas. Easy Buttons are intensely personal products meant to help you streamline what you do at your computer every day. It may be formal communication, document creation, marketing content creation, data analysis, legal review, planning, and much more.

You likely have some ideas already, based on the list you created in the last chapter. Starting with your raw list, follow the steps below and start working on your portfolio.

1. *Simplify and consolidate your list.* If you followed all our brainstorming tips, you probably have plenty of overlapping challenges and solutions on your list. You'll need a list you can prioritize and work through, so spend some time (we recommend having your LLM help you) cleaning it up until you have a single list with all, or mostly all, unique ideas.

2. *Categorize and analyze your list.* We recommend putting your list into a spreadsheet and adding some columns for "category," "impact," and "effort." Category is the aspect of your work impacted. Impact should be a number that reflects the value to you, your role, your team, or your company (maybe 1 to 10). Effort is the complexity of the task for an LLM to accomplish (i.e., how many things the prompt will have to handle at once and how many steps it might require to do successfully). Again, work with your LLM to help you iteratively fill in these fields, particularly the effort field.

3. *At long last, prioritize your list.* We realize you were already tempted to start sorting it, but we wanted you to wait until

you had some data in those fields. Once you do, we recommend sorting by effort and complexity first and then looking to see which items float to the top as the easiest to do with the biggest impact. These are the big-ticket challenges ripe for turning into Easy Buttons.

4. *Add to your portfolio.* Now that you have a better idea of which aspects to focus on, you're free to start building more Easy Buttons. Flip back to Chapter 3, Step 3, and you'll find the 4-step approach that includes making sure you have a simple task, working with your LLM to work through building the prompt you need, iteratively testing and updating it, and then making it easily accessible so you can use it often.

5. *Make your shortcuts "the way."* Figure out ways to make your Easy Buttons part of your daily work schedule. Habits die hard, and it'll be tempting to go back to your previous methods. Keep them handy and keep using them, and you'll continue to see the benefits of your work here.

6. *Keep checking on technology.* As generative AI models get smarter and smarter, you'll want to take advantage of improvements, especially when they will make a difference for your Easy Buttons. Just be careful not to invest more time in redesign than you'll get back, because sometimes the improvements are just tweaks and won't help as much as you think.

Unlock Easy Button Chains

Remember when we suggested you add a "category" column to your list of challenges? The sneaky reason for this is that we wanted you to be able to see the linkages between potential Easy Buttons. If you lean toward being a centaur (wanting to make your LLM do a lot more), you might start feeling excited here. Because the further you go down the path of building your Easy Button portfolio, the sooner you'll see them starting to group together, to the point where the results of one can feed into another… and another.

To show what we mean, we'll illustrate with an Easy Button Chain that creates a weekly accomplishments report. One Easy

Button would do poorly because there are multiple steps here. But an Easy Button Chain could include the following series of buttons: (1) interview you for ideas and their impacts (Note: this can be done out loud in some LLMs, so you might not even have to use a keyboard), (2) generate a single list of accomplishments in your tone and style, (3) group accomplishments into categories your boss needs, (4) create a simple summary narrative of everything you did, and (5) output a document with the summary and list in one place. Asking your LLM to build this would take a series of prompts and be difficult to do in one shot, but these 5 Easy Buttons put together in a chain can reduce this 2+ hour effort down to 15 minutes and dramatically improve the quality and consistency of your report.

The secret is taking the time to break down your process into tasks that can be automated, because these may be much different than the instructions you would give to a human. You might have to think about it as if you're going to have a new intern follow the steps every time they do them. Meaning, they have zero clue how to create the final product, but they can follow the step-by-step instructions you create.

While each Easy Button Chain will inherently be unique (to you and to your process), the five steps below are a good framework to help you create them successfully.

1. *Pick the right product or process to automate.* At first, we recommend adding Easy Buttons before or after one that already exists. For example, if you're already using one to interview accomplishments from yourself each week, maybe you can add the new buttons that will complete the full process, including the email you normally write to your boss. After that, you can start sorting by those categories you created in your challenges list and seeing what processes or products are simple enough and consistent enough to warrant a deeper dive into automation.

2. *Create a detailed breakdown of the steps you take.* Remember, you are explaining this to a clueless intern, so it has to be micro-management level detailed. In most cases, this will be

significantly more detailed than the steps listed in your company's process guide. We recommend starting with the smallest possible step (e.g., "write a sentence using this pattern and style") to ensure your LLM can consistently meet each requirement. You can always combine them later if the technology can handle it.

3. *Create all the Easy Buttons in your chain.* Take your time and follow the steps for each Easy Button, building one at a time and making sure they produce the consistent quality you need. This is the shortest step but by far the hardest!

4. *Compile the results into an Easy Button Chain.* Put together a document that is the new workflow for your dummy intern. Lay out all the Easy Buttons in order. For each, include instructions on what types of inputs the system needs and what type of output is expected (prompts don't always give you exactly the same response, after all).

5. *Test, update, and repeat until foolproof.* Practice using your new Easy Button Chain a few times to make sure it does what you hoped. Consider even testing it in a different LLM to see how robust it is. Ask a colleague to try it and see if they get the same results, or if the instructions are clear. And make tweaks along the way as needed.

Yes, creating Easy Button Chains will take work. This is an investment of your time and effort to build something you can use over and over at work. But it can also be a game changer in managing your workload and upping the quality and consistency of the products you create.

In fact, we should really ask you to give us that five-star Amazon review now, after we've landed so much value right on your desk, and before you forget who gave you these great ideas! If not, it's okay… we can remind you again later. *small business sweating*

Custom GPTs, Claude Projects, and Gemini Gems

If you've started following AI news or talking to AI enthusiasts, you may have caught wind of custom automation tools built into your favorite LLM. Put simply, these are built-in ways for your tool to help you with your Easy Buttons and Easy Button Chains. We see them as the next level up in building repeatable, sharable automation. Plus, they're kind of fun to play with.

So what are they? The best way to understand them is to start trying them out, but we'll give you the quick overview and differences here. At their core, both are essentially prompts that are fixed in place, with your documents and instructions already embedded and tweaked so that you can just hit "go," and it will do what you want. Plus, you can go straight to them with hyperlinks on your LLM dashboard (or favorites in your browser)—eliminating the need for copy-pasting prompts from your Easy Button portfolio.

But they are not perfect and not created equal. As of the writing of this book, here's our take:

- Custom GPTs are very easy to create and modify, and they're also easily sharable within a GPT Teams license or in the GPT Library to others. You can also string together GPTs in a single chat, allowing them to be their own kind of Easy Button Chain. However, they are only reliable for consistently performing small or relatively simple tasks, so we find they often fall short of the mark when it comes to high-complexity work products.

- Claude Projects are much more powerful, handling multiple Easy Buttons in one place and delivering better products than ChatGPT. However, they are less intuitive to create, don't offer a preview/testing option, and are only shared within a Teams group license.

- Gemini Gems have the advantage of integrating directly into your Google Workspace (Gmail, Google Docs, etc.), making it a powerful tool for working with data you maintain within that ecosystem. However, Gemini is still evolving, Gems may not follow directions as closely as you might prefer, and you can't embed files within a Gem for repeat access to common background materials.

Our recommendation is to create a few of your most-used Easy Buttons using GPTs, Projects, or Gems and see how far you can take them. There's definitely a lot of value here, and we don't want you to miss out on it. Below are a couple of links to simple training videos to get you started on using these tools:

- GPTs: https://www.youtube.com/watch?v=ABVwhZWg1Uk

- Projects: https://www.youtube.com/watch?v=lWcyLDoCHsw

- Gems: https://www.youtube.com/watch?v=WoB3iY-sWfc

And no, we're not affiliated with any of these content creators. If these don't work for you, keep searching around on YouTube and many others will come up!

As you implement these more complex solutions, remember to follow the same Easy Button steps we keep recommending. For now, we also suggest you keep maintaining your Easy Button portfolio. That way, you can keep working even if one LLM is busy or producing poor results one day.

Tailored Technologies

If you haven't already felt it, these Easy Buttons and Easy Button Chains are essentially a new way of programming that LLMs have made accessible to everyone, because they use natural language as their coding language. Yes, you can officially tell your friends that you're a freelance software developer. And professional and freelance software developers are taking this much further than we've described. This is because tools like Claude and ChatGPT have interfaces that can be used externally, and savvy developers can pull these capabilities into their own systems to build much more powerful tools.

We referred to these tools previously as Tailored Technologies, and we see them as the next echelon of automation for you to consider. These tools are pre-built Easy Button Chains with features designed into them to make the interface easier and the results more consistent and accurate, all without the need for you to keep managing and updating your Easy Button portfolio. And they're popping up everywhere, for almost every industry and type of job.

The downsides are that these tools cost money (a different kind of investment to talk to your boss about), it's often difficult to determine how your company's proprietary information will be protected, and it's unclear how they will evolve to accommodate the rapidly evolving technology they rely so heavily on. They are also brand new, so they typically have bugs to work out, are constantly being updated, and may not even be available next year. Plus, there are so many options out there that the process of narrowing them down to the one that will help you the most is just as time-consuming as building your own.

We're not saying they aren't valuable, but they are certainly bigger investments. As such, they'll take more careful study and discussions with stakeholders to weigh the pros and cons of different options. For now, we think it's important to be aware that they exist and to keep an eye out for those that will help you and your team boost productivity to yet another level. We've also included a link to our website in the "Thank You" section, where we post our favorite sources for sorting through some of these tools.

PART 5: HOW CAN I PROTECT MY CAREER?

Way back in the introduction to this book, you might remember wondering if you were a laggard or how you fell in Dr. Rogers' scale of technology adopters. This hopefully motivated you to press forward into the book to prove yourself wrong, but now we need to dig deeper. As technology continues to advance at the accelerating pace we've been seeing, we want folks who read this book to be the ones employers are desperately seeking. Not just because they understand how to use AI but because their personality (and hopefully resume) screams "experimenter," "adaptor," "innovator," and "general challenge facer."

So, how do you get there?

First, it's important to understand that we all have a little Laggard in us, just as we all have a little Innovator. You may be perfectly willing to test out a new recipe or learn a new language, but you can't even begin to wrap your mind around Excel pivot tables. The goals of Part 5 are to help you start cultivating a continuous growth and continuous learning mindset for yourself, building the mental muscles of an Innovator, positioning yourself at work to be an AI Advocate, and generally learning to embrace new technology. This is a long path and truly a lifelong process, but you can't get anywhere sitting still.

You Might Be a Laggard If…

Before she got her Ph.D. back in the early 1970s, Dr. Carol Dweck was watching her Yale University colleague Dr. Martin Seligman study how animals and humans respond to uncontrollable negative events, and she couldn't help but be curious about how some people just don't seem to give up. She started working with local elementary school children, giving them puzzles to solve beyond their ability and seeing how they reacted to failure. She found that some kids would give up, while others took to the challenge like Paul Atreides to a stillsuit on Arrakis (sorry for the deep Dune reference, it just felt right).

Her theory was that children have fundamentally different views of their abilities. They either believe their intelligence is fixed or that they can learn and grow. Like any good scientist, she then spent the next several decades developing this theory into a deep understanding of fixed and growth mindsets, and the complex spectrum of how we develop and apply these mindsets to different aspects of our lives. She put most of this research together in her groundbreaking, bestselling work "Mindset: The New Psychology of Success," which shows how dedication and effort can lead to greater success in any aspect of your life.[13]

Okay, enough of the history lesson. How does this help me get better at using AI?

A note of caution before we proceed: We're not psychologists and won't claim to be. And we know the field of psychology is always evolving. But we can still try to learn and learn to try, can't we? No

---

[13] Dweck, C. S. (2006). Mindset: The new psychology of success. Random House.

room for Laggards on this moving train!

Now, the first thing you need to understand is where you are on the mindset spectrum for situations like adopting new technology. It won't surprise anyone to find out that most Early Adopters of generative AI must absolutely be growth mindset people. It's too new a thing, and there are still plenty of excuses to avoid it altogether. And unless you're being forced to read this book by your boss (if so, a big thanks from Frontier Academy to that wonderful person), buying this book puts you in a good starting frame of mind.

Fortunately, figuring out where you land on the scale of adopters (from Laggard to Innovator) is a perfect generative AI discussion! We dropped a link into the "Thank You" section of this book, and we'll put one here ([14]) as a cited reference in case you're on a Kindle and want to try it right away without losing your spot.

That said, we still recommend going through the process of talking to AI. Just ask your favorite chatbot if it's familiar with Dweck's book on Mindset, then maybe try a "back and forth interview" followed by an assessment and validation discussion to see what you learn. As far as continuing your learning journey in this book, make sure you are focused specifically on your mindset as it relates to new technology at work, the use of generative AI, or something similar. Finding out you have a fixed mindset against board games could be interesting, but it might not help you on this particular journey.

Here are a few broad questions you might try asking yourself (or having generative AI ask you) as you work to figure this out:

- How do you react when a project hits a snag? Do you get frustrated and want to give up, or do you see it as an opportunity to learn and improve?

- When you need to put in extra effort, do you see it as a sign that you're not good enough or as a necessary part of the growth process?

---

[14] https://chatgpt.com/g/g-scpbGNlOY-laggard-to-innovator-assessment

- How do you feel when you receive constructive criticism? Defensive or grateful for the feedback?

- When a colleague succeeds, do you feel threatened or inspired?

- How do you approach new challenges? With excitement or anxiety?

Once you achieve some modest amount of self-awareness on this topic, we think it will help you tremendously in plotting a solid path out of the Laggard swamp. Next, and we mean this regardless of where you learn you stand, you need to make a conscious decision to continue to develop and strengthen your growth mindset. This is a continuous process that we can all learn to do better. Below are some tips to consider that can help with this. Some are simple, and some are daunting, but all will flex and stretch the mental muscles you need to be the Innovator you're trying to be.

- **Embrace the power of "yet":** When you catch yourself saying "I can't do this," add "yet" to the end of the sentence. This simple word opens up possibilities and reminds you that skills can be developed.

- **Set learning goals:** Instead of focusing solely on performance goals, set specific learning goals for each project or quarter. For example, "I want to improve my data visualization skills" alongside "I want to deliver this report."

- **Seek out challenges:** Regularly volunteer for projects that stretch your abilities. It might feel uncomfortable, but that's where growth happens.

- **Cultivate a feedback-seeking mindset:** Actively ask for feedback from colleagues and superiors. When you receive constructive criticism, resist the urge to become defensive. Instead, thank the person and ask for specific ways you can improve.

- **Celebrate effort and progress:** At the end of each week, reflect on the effort you've put in and the progress you've made, regardless of the outcomes. This helps shift focus from pure results to the process of growth.

- **Reframe failures:** When things don't go as planned, ask yourself, "What can I learn from this?" Turn setbacks into valuable data points for future success.

- **Develop a learning ritual:** Set aside time each week for learning. This could be reading industry publications, taking an online course, or practicing a new skill.

- **Use growth mindset language:** Pay attention to your self-talk and the way you communicate with others. Use phrases that emphasize growth, learning, and effort.

- **Seek out growth-minded mentors:** Surround yourself with people who embody a growth mindset. Their attitudes can be contagious.

- **Practice self-compassion:** Remember that developing a growth mindset is itself a growth process. Be patient with yourself as you work on changing long-held beliefs and habits.

Remember, developing a growth mindset is an ongoing journey. It requires conscious effort and regular reflection. But as you persist, you'll likely find yourself becoming more resilient, adaptable, and open to the continuous learning that today's fast-paced work environment demands. By embracing a growth mindset, you're not just keeping up with the pace of change in your industry - you're positioning yourself to thrive in it. Your future self will thank you.

Get Out Over Your Skis

Typically, the phrase "be careful not to get out over your skis" is used as a warning to move slowly. Leaning forward on snow skis is the best way to speed up and potentially lose control. In this case, however, we're advocating speed. Think of the mountain as a sliding avalanche rather than a static, gentle slope. When the mountain is moving beneath you, the best way to survive is to keep up with it.

So, how do you keep up with it?

Below are some habits that will help. If the list is too daunting, maybe just pick a couple and start trying to add them to your work routine. You won't go any faster standing still.

- **Stay Informed:** Regularly read industry publications, attend conferences, and participate in professional networks to stay aware of emerging technologies.

- **Keep Learning:** Set aside time for learning new skills. Platforms like Coursera, edX, and LinkedIn Learning offer courses on various technologies.

- **Experiment:** Allocate time to experiment with new tools and technologies, even if they're not immediately applicable to your current work.

- **Collaborate with Others:** Work with colleagues from different departments to gain diverse perspectives on how new technologies can be applied.

- **Surround Yourself with Smart People:** Pair up with colleagues who may be more familiar with emerging technologies.

- **Build up a Personal Learning Network:** Develop a network of professionals, both online and offline, to share knowledge and experiences.

- **Fail Forward:** Nurture a mindset of accepting and embracing failure as a learning opportunity.

- **Use Baby Steps:** Find small ways to change your process or test new tools, gradually increasing complexity as you become more comfortable.

- **Teach:** Teaching others what you've learned is a great way to reinforce your own understanding and help create a culture of continuous learning in your workplace.

- **Reflect and Adapt:** Regularly reflect on your learning process and adapt your strategies as needed.

Close Your Eyes, Raise Your Hand

Have you ever worked in a position where you completely ran out of work to do? Neither have we. And now business leaders have an entirely new challenge: what to do with this AI thing?

This is where the opportunity to position yourself in your organization as an AI advocate, as someone with an interest, passion, and enough information to start trying some things out. If your imposter syndrome just lifted a few of your neck hairs, remember that this generative AI thing is continually new and evolving. Meaning there's no such thing as a "proven expert." There are only the triers and the didn't-try-ers.

In previous sections, we worked through some specific ways to add value at work, including identifying challenges and solutions to them. All of these have been helping develop your growth mindset muscles, so let's try to flex them a bit. Let's figure out how you can put yourself in a position that requires learning and growth. Especially if it's also an opportunity for failure.

How specifically? Volunteer. Walk up to your manager and offer to solve a challenge for them. Offer to test out a new tool, use AI to take on a challenge no one else has been willing to, or give training to your peers on a subject everyone needs help with. You might already have an idea in the back of your mind, the one you've been dreading thinking about because of the discomfort it gives you. If not, here are a few discussions you can have with generative AI to spot and even flesh out some ideas on what exactly to sign up for:

- Engage in a dialogue about your company's current challenges, digging deeper into root causes and exploring unconventional solutions you might spearhead.

- Have it analyze your job description in relation to relevant industry trends. This can highlight areas where your expertise falls short, inspiring you to propose leading initiatives in these areas.

- Have a conversation about your industry's future, envisioning upcoming changes. This could reveal opportunities to pioneer new technologies or processes within your organization.

- Have it analyze your company's structure and discuss potential interdepartmental collaborations. This might suggest innovative projects that require you to step outside your usual domain.

- Engage in a free-flowing ideation session about your product or service. Generate out-of-the-box concepts, potentially inspiring you to volunteer to develop a new offering.

- Have a detailed conversation about your company's current processes. Explore automation potential using generative AI or other technology, potentially inspiring you to lead a major transformation project.

- Discuss various leadership approaches and your company culture. This might reveal leadership gaps you could fill by proposing new initiatives or programs.

- Engage in a detailed analysis of your customer experience. Uncover pain points, leading you to propose leading a customer-centric overhaul project.

- Even if not immediately relevant, discuss potential global expansion. The insights gained might inspire you to volunteer to research and plan new market entries.

Once you pick yourself a nice, uncomfortable task to work on, get yourself permission to proceed from whoever you need to. Once you have that, have an ongoing back-and-forth conversation about the task or challenge, where to start, what tools you might need, what you need to learn, training you should look into, a book you can read, or any other advice you need. During the project, remember to ask AI for help overcoming pesky pop-up challenges and refining your plans when needed. You can even ask for encouragement along the way as you work!

And remember, the point is the learning, the growth, and the mental muscles you'll need to thrive in this technology revolution. You're not just trying to find a spectacular way to fail or a great story to tell your grandkids. You're trying to drag your imposter syndrome, kicking and screaming, out into the light where it can't survive. You're trying to become a future-proof employee.

PART 6: WHERE IS ALL THIS HEADING?

QUICK   *Ask AI to describe your job as it might look in 2030. This peek into*
WIN   *potential futures can help you spot opportunities to prepare today.*

Our tips, tricks, and advice have been pretty practical so far, wouldn't you agree? That was the goal, at least!

Now, though, we want to take you on a thought journey beyond the horizon of the known to a place of speculation and guesswork. The hope is that we give you something very difficult to find when you're buried in the day-to-day grind of learning to be an Early Adopter and an Innovator. Namely, perspective. A vision of the possible future to help you guide today's plans.

A word of caution here. First, we're likely to be wildly wrong. Even though it seems logical today, things are changing fast enough for that logic to fall apart quickly tomorrow. Second, these views are optimistic. Even if some aspect of the future looks like we said, there will likely be plenty of difficult days between this day and that one. Third, and most importantly, it's out of your control. Direct your anxiety and hand-wringing elsewhere. If you can't, take a deep breath and read this book again with renewed motivation and urgency.

Anyway, please secure any loose items and remember to keep your arms and hands inside the cart at all times.

Hard Right Turns

As generative AI and a host of other technologies continue to improve, companies of the future might be able to turn on a dime. We like to picture the future corporate landscape like a field trip of kindergarteners at a skating rink, only they all have office chairs and fire extinguishers. Companies will be able to respond at blazing speed to market changes, almost automatically shifting skillsets and supply chains to use new tools, follow new processes, and work on new projects.

As generative AI makes everyone in the company able to perform basically any job, companies will have an infinitely flexible pool of capabilities to draw from. As communication and planning mechanisms improve (as they already are, and fast), the inefficiencies between executives and supervisors and performers will dwindle into nothing, to the point where a new company strategy can be planned and rolled out to individuals in days instead of months or years. Processes, process owners, and process management will be a thing of the past as automated training tailored to specific plans rolls out right alongside assignments. Further, automation in robotics and supply chains will mean the pipeline of supplies, parts, or products will be increasingly "on demand," enabling quick turn shifts to accommodate the executive plans.

The impact on knowledge workers will be that corporations place an increasingly high value on adaptability, continuous learning, and a willingness to work in a new area with new tools, with limited training, and without a process net. The assembly line days, where everyone stays in their lane and follows detailed instructions, will decay into an interesting historical anecdote.

No Resume Needed

"Shapeshifters" are the knowledge workers of the future. They can jump into any job and use AI to successfully accomplish it. It should be obvious, but this book has been preparing you for exactly this future by helping you learn to use today's tools and build a mindset of continuous learning. Once employers start to see the value of these individuals, we predict an increasingly wide divide between those who embrace the AI tools that make this possible… and those who do not.

One interesting side effect of the existence of Shapeshifters is that the value of the traditional resume could fall drastically, even vanishing altogether. Once upon a time, we used to memorize the phone numbers of anyone we wanted to talk to regularly. Today, the knowledge of how to connect to a specific person is stored in our mobile devices, so we no longer need to learn the numbers. We see a similar thing happening in the workplace. As generative AI gives everyone access to all knowledge, someone's education and experience in a specific topic could actually represent a fixed mindset that has less value to an organization.

Taking this one step further, large companies aren't going to be the only ones that see this. Piles of one-to-three-person companies will spring up and be able to accomplish anything they set their mind(s) to. The already-growing gig economy will flourish, offering Shapeshifters-for-hire that can plug into any project for the time needed and then disconnect just as quickly. The prices of consultants will go down along with the need for full-time employees. And for individuals with these skills, it will mean and increasingly high degree of freedom – to work when needed and relax in between.

Interest-Based Culture

When asked today, "What should my kids be studying?" our answer is that they should be trying to figure out what they most enjoy doing.

Looking inside companies, especially ones that keep large workforces in place, we think a new "interest-based culture" will emerge where employees are assigned to the projects they are most interested in. Your Talent Management Team will review your "passion profile" and place you in areas that will maximize your engagement and value to the company (where you'll have the most fun and impact). Or a Talent Pipeline Team from another company might have spotted your profile and thought it would fit well for their next project. In fact, keeping your passion profile up-to-date will be the next challenge as you try new things and learn what your next area of interest might be.

The point is that you should start to open your aperture and look for things that interest and motivate you. Because even in a world where AI Agents can do just about anything, we like to believe that passionate humans will continue to add a ton of value.

The Skills That'll Matter

It's hard to predict how skill needs will change, but there are some clear trends today that suggest that meta-skills will be the job currency of the near-ish future. Yes, we mean in place of education and even experience unless that experience further proves your meta-skill proficiency. In fact, Hiring Lab (funded by Indeed) studied job postings over time and found that "mentions of college degrees have fallen since 2019 in 87% of occupational groups."[15]

If you haven't heard the term before, meta-skills are broad mindset-related capabilities that help you excel in a wide variety of situations using AI tools and technology. Communication, flexibility, adaptability, critical thinking, and even emotional intelligence are all examples. The term "prompt engineering" was thrown around heavily for a while, but that's really just the ability to communicate more effectively and logically. But if you've been dabbling with some AI tools, you probably realize that you're getting better and better at getting what you need out of it.

As an aside, younger generations who grow up with generative AI apps on their phone are already learning some of these skills by accident. As an example, Character.ai is an entertainment platform (mostly) where you can chat with a powerful LLM in the form of a host of different personalities and voices. Of the 200+ million

---

[15] https://www.hiringlab.org/2024/02/27/educational-requirements-job-postings/

people using this tool, 60% are between 18 and 24 years old.[16]

Of course, there will still be jobs that simply require humans—not because AI can't do the job, but because humans sometimes just like talking to other humans. The 2024 Restaurant Technology Landscape Report showed that only 33% of adults are currently in favor of talking to AI to place orders at a restaurant.[17] More than that, we're predicting that humans will eventually be involved in the end-to-end customer experience, allowing customers to opt-out of chatbot answers, talk to real sales people, and even order food from an actual server.

So, how do you start building the skills you need for a future workplace? Having this book and being a human are two great steps in the right direction! Keep working on being the kind of downhill skier that can handle avalanches, or at least the kind of employee that can roll with fast-moving change. Keep learning to communicate with AI tools, particularly as they evolve, so you can get the most out of them at every stage of your career. And keep being human!

Further Down the Rabbit Hole

So, what does it even mean to be human? Interestingly enough, this may be changing too. The CEO of NVIDIA famously said in 2023 that "for the very first time in our history, in human history, biology has the opportunity to be engineering, not science. When something becomes engineering, not science, it becomes less sporadic and exponentially improving."[18] Futurist Ray Kurzweil, who has one of the most remarkable prediction records at roughly 86% accuracy, refers to "people with cloud-connected neocortices" and "radically new possibilities for how the brain itself processes experiences" in his 2024 book, *The Singularity is Nearer*. We may be looking at bio-engineered enhanced humans, genetic engineering

---

[16] https://whatsthebigdata.com/character-ai-statistics/

[17] https://go.restaurant.org/rs/078-ZLA-461/images/NatRestAssoc_TechLandscapeReport_2024.pdf

[18] https://www.youtube.com/live/9hzVdV63scU?si=2tyQz5XF8HsZxv-c&t=2956

that allows us to live forever, and brain control interfaces that change how we interact with the world around us entirely.

Told you we'd be going further down the rabbit hole.

Tomorrow's tools will also be exciting. Keep your ears open for the word "agentic," as that may mean generative AI tools are starting to race ahead of us, anticipate our next steps, and take action on our behalf. Watch also for advances in extended reality systems that allow high-quality virtual collaboration and double down on the post-pandemic enthusiasm for remote work. If you don't already feel like you're living in the future, brace yourself.

In fact, as humans and technology continue to evolve and mature, we may increasingly look backward to this time as the Slow Age, before the sky really opened up to us and let us accomplish anything with only the thoughts in our minds.

PART 7: WHAT'S NEXT?

QUICK WIN

Create a simple AI action plan by asking: "What's one small way I can use AI in my work tomorrow?" Then commit to trying it – your future self will thank you.

As we reach the end of our journey through the world of generative AI, let's take a moment to reflect on the key takeaways and strategies we've explored together. This book has been more than just an introduction to AI tools; it's been a roadmap for revolutionizing your personal and work life for what the future holds.

We began by demystifying AI, breaking down complex concepts into digestible pieces. Remember our discussion on Large Language Models and how they're transforming industries? We explored why not to worry about the robot army, why your boss feels nervous, and what you can do today to start protecting your job.

We then dove into practical applications, giving you some tips and best practices for working with generative AI in your personal and work life. From creating your first Easy Button to starting to build your own custom automation, we've covered a wide range of strategies to boost your productivity and effectiveness.

If you followed along, took our advice, and started working

through your very own existential crisis (or crises, maybe), you've been on quite the journey. With any luck, you've come out of the other side at least understanding some of the basics, if not starting to feel comfortable working with AI for a handful of tasks. Yes, there might be a learning curve, and yes, it might feel uncomfortable at first. But hey, every expert was once a beginner. So wherever you are in your learning process, keep learning, keep trying, and keep learning forward. Heck, even put it on your resume, "AI Badass!"

From Imposter Syndrome to AI Superstar

Think back to Sarah, our protagonist from the beginning of the book. When we first met her, she was grappling with imposter syndrome, unsure of her place in a rapidly evolving workplace. By taking that first step and integrating generative AI into her work, she not only conquered her self-doubt but revolutionized her role. Sarah's story exemplifies what we've been working towards throughout this book: The Death of Imposter Syndrome.

You're now one step closer to that reality. By making it through this book, you've armed yourself with the knowledge and strategies to follow in Sarah's footsteps. You understand the potential of generative AI, you have some clear methods to make it work for you, and you're ready to start using it more and more. In essence, you've been given the tools to slay the dragon of self-doubt.

Beyond the Laggard

Think back also to Dr. Rogers' adoption categories we discussed earlier. When you first picked up this book, you might have seen yourself as a Laggard, hesitant to adopt new technologies. But by making it to this final page, you've already moved beyond that.

This shift isn't just about using new tools. It's about adopting a new mindset - one of continuous learning, adaptability, and innovation. Whether you now see yourself as an Early Adopter, eager to implement new AI tools in your work, or even an Innovator, ready to push the boundaries of what's possible with AI, you're no longer watching from the sidelines - you're in the game, taking charge of your future and even the future of your industry.

And if you don't feel quite so confident, go ahead and take our free "Laggard to Innovator Assessment" GPT to find out. The link is under the "Thank You" section at the end of the book.

Continuing Your AI Journey

Remember, this book is just the beginning. The AI landscape will continue to evolve, and so should your skills. Stay curious, keep learning, and don't be afraid to experiment. Every interaction with generative AI is an opportunity to refine your approach and unlock new possibilities.

Consider joining AI-focused professional groups or online communities. Attend webinars and conferences on AI in your industry. Set up news alerts to stay updated on the latest AI developments. The more you engage with the AI community, the more opportunities you'll discover. Our website has some places you can start, and we dropped a link to that page in the "Thank You" section along with the others.

Finally, if you've found value in this book, we'd be thrilled if you could share your thoughts in an Amazon review. And here's a fun idea – why not use one of the generative AI tools we've discussed to help draft your review? We promise we won't be offended. In fact, we'd be delighted.

The next chapter of your generative AI life starts now! The future is AI-augmented, and you're now ready to thrive in it. Embrace the change and enjoy the journey ahead!

THANK YOU

Below are some links to help you get started as a small token of our appreciation for your purchase of this book (and for that great Amazon review you almost forgot about).

Laggard to Innovator Assessment

Click below to use our free Custom GPT that asks you a series of questions to help determine where you are on the scale from Laggard to Innovator. Feel free to use this often as you continue your growth journey and need some helpful advice on what to try next.

https://chatgpt.com/g/g-sepbGNlOY-laggard-to-innovator-assessment

Easy Button Designer

Click below to use our free Custom GPT that helps you turn your idea for an Easy Button into reality. It asks you questions to better understand your goal and then helps you create a prompt that aligns with our prompt creation recommendations.

https://chatgpt.com/g/g-6765d611af448191
8dfc54a12d177627-easy-button-designer

ABOUT THE AUTHORS

Dina Alkhateeb is a forward-thinking leader and expert in corporate communications, known for her ability to leverage innovative technologies to drive success. As an early adopter of emerging technologies, Dina has always been at the cutting edge of integrating new tools and methodologies to enhance workflow efficiency. Her career has been marked by a relentless pursuit of excellence and a productivity mindset that has consistently delivered results in fast-paced, high-stakes environments. Dina's expertise in developing high-performance teams and her commitment to individual growth make her an invaluable resource for professionals seeking to upskill in AI. Her practical approach and productivity mindset offer readers actionable strategies to effectively incorporate AI tools into their daily work, helping them not just adapt but thrive in the age of artificial intelligence.

Raymond Vogel is a seasoned professional with over 20 years of experience as a business leader, innovator, communicator, and engineer. Throughout his career, Raymond has been a catalyst for change, consistently seeking out and implementing cutting-edge technologies to drive efficiency and growth. His unique background, spanning from engineering automation to entrepreneurship, gives him a multifaceted perspective on the practical applications of AI in various professional contexts. Raymond's passion for continuous improvement and his talent for translating complex technical concepts into accessible language make him an ideal guide for individuals looking to harness the power of AI tools in their careers. His insights help readers navigate the AI landscape with confidence, empowering them to stay competitive in an increasingly AI-driven workplace.